AF688006

WAP's Intricacies

A Deep Dive Into Network Security's Architectures, Strategies, And Emerging Trends

Ethan Langley Bennett,Prescott

Copyright © 2025 by Ethan Langley Bennett,Prescott
All rights are reserved.

Table of Contents

1 Introduction to WAP Security

Wireless Access Points (WAPs) are the gateways to your network, and securing them is paramount. Understanding their fundamental architecture—from the radio frequency transmissions to the network interface card (NIC) and the associated control protocols—is the first step. A WAP essentially acts as a bridge, translating wireless signals into wired network data, making it a critical vulnerability if improperly secured. Think of it as the front door to your digital home; if that door is unlocked, your entire house is at risk. Neglecting WAP security exposes your network to a myriad of threats, from simple eavesdropping to sophisticated data breaches.

Robust security measures are essential, starting with choosing the right encryption protocol. Many still use outdated protocols, leaving their networks exposed. The transition to WPA3 is crucial, as it incorporates significantly stronger encryption methods and addresses many of the weaknesses inherent in its predecessors. Failing to update poses significant risks, particularly in sensitive environments. While upgrading might seem technically daunting, the long-term protection outweighs the short-term inconvenience. Consider it an investment in the security of your valuable data. Remember, even seemingly minor security oversights can have massive repercussions.

Network segmentation is your next line of defense. Imagine your network as a large building divided into smaller offices, each with its own security system. VLANs (Virtual Local Area Networks) allow you to logically segment your network, isolating different sections and preventing unauthorized access. For instance, separating your guest network from your internal network drastically reduces the risk of a breach affecting sensitive data. Combine VLANs with strategically placed firewalls to create a layered defense system, effectively creating a secure perimeter around valuable data.

Beyond basic protocol implementation and network segmentation, consider advanced encryption. AES (Advanced Encryption Standard), with its various key lengths, offers robust data protection, effectively scrambling your data to render it incomprehensible to unauthorized parties. However, the strength of AES relies heavily on secure key management; weak keys negate the strength of the encryption itself. A robust PKI (Public Key Infrastructure) ensures secure key exchange and distribution, reducing the risk of compromise. Furthermore, implementing Perfect Forward Secrecy (PFS) ensures that even if a key is compromised, past communications remain secure, thus safeguarding against past data breaches. Elliptic Curve Cryptography (ECC), known for its efficiency and security, is increasingly important in resource-constrained environments like mobile devices and IoT devices connected to your WAP.

Proactive security measures are crucial in mitigating threats before they can materialize. Employing Intrusion Detection/Prevention Systems (IDS/IPS) acts as a real-time security guard, continuously monitoring network traffic for suspicious activity. IDS passively monitors and reports anomalies, while IPS actively blocks malicious traffic. Integrating these systems with a Security Information and Event Management (SIEM) system offers a centralized view of security events, allowing you to identify trends and respond swiftly to potential threats. Remember, even with robust security measures, regular monitoring and analysis are essential to maintaining a secure environment.

Regular security audits are vital for assessing the effectiveness of your security measures. Vulnerability assessments identify weaknesses in your system, allowing you to address them before malicious actors can exploit them. Penetration testing simulates real-world attacks, offering a practical understanding of your network's security posture. By actively seeking out and mitigating vulnerabilities, you greatly enhance your network's overall resilience. This proactive approach minimizes your risk profile and improves your defenses. Consider these audits not as an expense, but an investment in preventing future financial and reputational damage.

The future of WAP security is constantly evolving. Emerging trends such as AI-driven threat detection, next-generation encryption algorithms, improved IoT security integration, and quantum-resistant cryptography are already shaping the landscape. Staying informed about these advancements is critical for maintaining a robust security posture. The integration of zero-trust architectures, verifying every user and device before granting access regardless of location, represents a fundamental shift in security paradigms.

Embracing these advancements is crucial to safeguard your network against both current and future threats. Continuous learning and adaptation are critical to effectively managing the ever-changing security landscape.

1.1 WAP Fundamentals

Wireless Access Points (WAPs) are the foundational elements of any wireless network. Understanding their fundamental architecture is paramount to securing them effectively. Think of a WAP as a translator, converting the signals from your wireless devices into a format your wired network understands, and vice-versa. This process involves several key components: the radio frequency (RF) transceiver, responsible for sending and receiving wireless signals; the antenna, which amplifies and directs those signals; and the network interface card (NIC), which connects the WAP to the wired network. Finally, a powerful processor manages these components, ensuring smooth operation and handling security protocols. A thorough grasp of this architecture underpins all subsequent security considerations.

Effective WAP security begins with a solid understanding of its operational capabilities. Beyond basic connectivity, WAPs offer features like Quality of Service (QoS) for prioritizing specific types of traffic, bandwidth management to control data flow, and access control lists (ACLs) to restrict access based on MAC addresses or IP addresses. However, relying solely on these built-in features is insufficient. Proper configuration is crucial. For example, disabling unnecessary services reduces the attack surface. Similarly, regularly updating firmware patches vulnerabilities exploited by malicious actors. These seemingly simple steps form the bedrock of a robust security strategy.

The physical placement of your WAPs significantly impacts network security. Optimal placement minimizes signal interference and dead zones, enhancing signal strength and reliability. Strategically positioning WAPs also helps contain the signal within a designated area, reducing the risk of unauthorized access. Consider environmental factors, such as building materials and obstacles that might weaken signal strength. Proper planning ensures consistent, high-quality connectivity while limiting the area exposed to potential attackers. Remember, a strong signal in a well-defined area is less likely to be intercepted by someone outside of that area.

Securing your WAP goes beyond the device itself. Network segmentation plays a crucial role. By dividing your network into smaller, isolated segments, you limit the impact of a potential breach. Imagine a scenario where one segment is compromised; if properly segmented, the attackers would have limited access to other sensitive data. Techniques such as Virtual LANs (VLANs) allow you to logically divide your network without needing physical changes. This allows for granular control over access and traffic flow, confining any potential security incident. Using VLANs in conjunction with carefully configured firewalls provides a layered defense strategy to restrict access across segments.

Effective security isn't just about technical measures; it demands consistent monitoring and maintenance. Regularly checking your WAP's logs helps identify anomalies indicative of malicious activity. Looking for unusual login attempts, high bandwidth usage spikes, and failed connections can give clues to potential problems. Moreover, implementing robust logging practices, securely storing logs, and regularly analyzing them can highlight patterns and trends in network traffic, helping predict and prevent future threats. These logs should be treated as crucial security data, and their protection should be prioritized. Don't leave them vulnerable.

Remember, effective WAP security is an ongoing process, not a one-time task. The threat landscape constantly evolves, demanding continuous vigilance. Staying informed on emerging threats and vulnerabilities, regularly reviewing your security measures, and adjusting your strategies accordingly are crucial for maintaining a strong security posture. Continuous education and awareness form an integral part of building resilient network security. Proactive monitoring, threat intelligence, and adaptation to the changing security landscape are vital for securing WAPs and wireless networks in the long term. Therefore, a vigilant and proactive approach to WAP security should be maintained, constantly adapting to emerging threats and best practices. The principles discussed here form the base for a comprehensive understanding of WAP security and how to ensure that your network remains protected in ever-changing environments.

1.2 Security Threats Overview

Wireless Access Points (WAPs), the ubiquitous gateways to our wireless networks, are unfortunately prime targets for a variety of attacks. Understanding these threats is crucial for implementing effective security measures. These aren't just theoretical vulnerabilities; they represent real-world risks with potentially devastating consequences, from data breaches costing businesses millions to complete network disruptions crippling operations.

Unauthorized access is a constant threat. Malicious actors might exploit weak passwords or default configurations to gain entry to your network, potentially accessing sensitive data or using your resources for illicit activities. This is especially problematic for public WAPs, where security is often overlooked in favor of convenience. Imagine the ramifications if a hacker gained access to a company's network via an unsecured guest Wi-Fi. The implications for data confidentiality, integrity, and availability are severe.

Eavesdropping, often facilitated by the open nature of wireless transmissions, poses a significant risk. Without robust encryption, sensitive data transmitted over the network, such as financial information or personal communications, is easily intercepted. Think about the consequences of an attacker listening in on your credit card details during an online transaction. The damage can extend beyond financial loss, impacting reputation and possibly leading to legal repercussions.

Denial-of-Service (DoS) attacks aim to disrupt network availability by flooding the WAP with traffic, making it unusable for legitimate users. This can range from simple attacks launched by individuals to sophisticated, distributed attacks that cripple entire networks. Consider a hospital network experiencing a DoS attack during a medical emergency—the consequences could be catastrophic. The disruption of essential services highlights the critical need for robust protection against these types of attacks.

Man-in-the-middle (MitM) attacks allow attackers to intercept and manipulate communication between two parties. By positioning themselves between the WAP and legitimate users, they can eavesdrop, modify data, or even inject malicious code. This kind of attack can compromise the integrity and confidentiality of data exchanged on the network. It's like a malicious third party intercepting a phone call; a MitM attack can insert false information or intercept sensitive conversations.

Rogue access points are unauthorized WAPs installed within a network without the owner's knowledge or consent. These can be used for malicious purposes, such as intercepting data, conducting surveillance, or launching attacks against other network devices. Consider the risk of an employee unknowingly connecting to a rogue access point, inadvertently compromising sensitive company data. Identifying and eliminating these rogue points requires a dedicated network security strategy.

Malware infections can compromise the WAP itself, turning it into a tool for further attacks. Infected WAPs can distribute malware to connected devices or act as a launching point for broader attacks against the network. Think about the ramifications of a compromised WAP silently distributing ransomware to every device on the network. This would cause significant financial losses and operational disruption.

Physical theft or tampering also presents a substantial threat. A physically compromised WAP can grant attackers full access to the network and its resources. Imagine a hacker stealing a WAP from a small business and accessing all the company's sensitive data. This could easily lead to a complete business collapse. Physical security measures must be implemented to protect the WAPs themselves.

Wireless network vulnerabilities constantly evolve, requiring proactive strategies to mitigate risk. Regular updates to the WAP's firmware, strong password policies, and the use of advanced encryption protocols are paramount. Ignoring these vulnerabilities increases the likelihood of a successful attack. Staying informed about the newest exploits and promptly addressing security flaws is a crucial aspect of WAP security. The cost of inaction far outweighs the cost of maintaining a robust security posture. Proactive, multi-layered security measures are needed to protect against the ever-evolving landscape of wireless threats. The potential cost of a breach in terms of data loss, financial penalties, and reputational damage is significant. Investing time and resources in robust WAP security is a wise and necessary choice.

2 Wireless Security Protocols

WPA/WPA2, the workhorses of Wi-Fi security for years, utilize the Temporal Key Integrity Protocol (TKIP) or Advanced Encryption Standard (AES) for data encryption. TKIP, while an improvement over its predecessor Wired Equivalent Privacy (WEP), is now considered obsolete due to significant vulnerabilities. AES, on the other hand, with its robust 128-bit or 256-bit key lengths, provides much stronger protection against unauthorized access and eavesdropping. However, even AES isn't impervious; implementation flaws and vulnerabilities in other parts of the system can still compromise security. Regular firmware updates are crucial to patching known vulnerabilities and ensuring ongoing protection. Understanding the differences and limitations of these protocols is essential for selecting the appropriate security level for a given network.

WPA3 represents a significant leap forward. It introduces Simultaneous Authentication of Equals (SAE), a more secure handshake process that replaces the vulnerable Pre-Shared Key (PSK) method used in WPA2. SAE employs a robust password-based key exchange resistant to dictionary attacks and offline cracking attempts. This dramatically enhances the security of the initial connection, a critical aspect frequently targeted by attackers. Furthermore,

WPA3 incorporates enhanced protection against dictionary attacks by using a stronger hashing algorithm and more complex key derivation functions. The improved authentication mechanism significantly reduces the risk of brute-force attacks, offering considerably better protection for your network. Adopting WPA3 is a crucial step in modernizing your Wi-Fi security infrastructure.

Open authentication, while seemingly convenient for public access points, exposes networks to substantial risks. Without any authentication mechanism, anyone within range can connect to the network, gaining access to all shared resources and potentially compromising sensitive data. Malicious actors can easily exploit this lack of security to launch various attacks, including man-in-the-middle attacks, denial-of-service attacks, and data breaches. The ease with which an attacker can gain access underscores the severity of this vulnerability. For public access points, implementing robust authentication, such as captive portals with user registration or strong password authentication, is essential to minimize risks and maintain a secure environment. Even in environments where open access might be desirable, a layered approach with additional security measures should be considered.

Choosing the right security protocol isn't merely about selecting the latest version; it involves understanding the specific needs and context of your network. A home network might have different security requirements than a large corporate network. The level of risk tolerance, the sensitivity of data being transmitted, and the technical expertise available all play vital roles in the decision-making process. A thorough risk assessment should be conducted before selecting a protocol to ensure that the chosen security

measures appropriately address the potential threats. Regular monitoring and auditing practices are also crucial to identify and promptly address any emerging vulnerabilities or security breaches.
Proper configuration is paramount, regardless of the chosen protocol. Default passwords should be changed immediately. Strong and unique passwords, ideally generated using a password manager, should be implemented. Regular password changes, as well as the implementation of multi-factor authentication whenever possible, contribute to a robust and secure network. Furthermore, keeping the firmware of your access point up-to-date is critical to patching security holes and protecting against emerging threats. Neglecting these basic best practices can severely compromise even the most advanced security protocols, rendering them ineffective against malicious attacks. Remember, security is a multifaceted process and requires attention to both software and user practices.

2.1 WPA/WPA2 Deep Dive

WPA and WPA2, while now considered legacy protocols, formed the backbone of Wi-Fi security for many years. Understanding their inner workings is crucial, even in the era of WPA3. Both protocols rely heavily on the Temporal Key Integrity Protocol (TKIP) or the Advanced Encryption Standard (AES) for data encryption, significantly bolstering the security of wireless networks compared to their predecessors, WEP. The fundamental difference lay primarily in their encryption methods and key management. TKIP, used in WPA, was a significant improvement over WEP but ultimately proved vulnerable to certain attacks. AES, on the other hand, provided far stronger encryption, becoming the preferred and eventually the standard method within WPA2. This shift showcased the rapid evolution of wireless security protocols, highlighting the constant need for improvement in the face of evolving threats.

The transition from WPA to WPA2 wasn't merely about upgrading encryption. It involved improvements in key management. WPA's key management, while better than WEP, still possessed certain weaknesses that were addressed in WPA2. This enhanced key management, combined with the superior strength of AES encryption, made WPA2 a considerably more resilient security protocol, significantly reducing the risk of successful attacks exploiting known vulnerabilities. The implementation of the four-way handshake in both protocols, however, remained a core element, providing a secure mechanism for establishing a shared secret key between the client and the access point. Understanding this handshake is crucial for grasping how authentication and encryption work within these protocols.

Despite its superior security compared to WPA and WEP, WPA2 wasn't without its vulnerabilities. The discovery and exploitation of the KRACK attack (Key Reinstallation Attack) significantly impacted the security posture of numerous WPA2 networks worldwide. This attack exploited weaknesses within the four-way handshake, allowing attackers to decrypt sensitive data. The vulnerability highlighted the critical importance of keeping software and firmware updated to patch known security flaws. The KRACK attack served as a stark reminder that even seemingly secure protocols require continuous monitoring, updates, and rigorous security audits to mitigate the evolving landscape of threats. It underscored the need for

proactive security measures rather than a purely reactive approach.
Therefore, a deep understanding of WPA and WPA2, including their strengths, weaknesses, and known vulnerabilities, is paramount. This knowledge allows network administrators to effectively assess the security posture of their networks. For instance, it helps determine whether a network still using WPA/WPA2 requires immediate upgrades to WPA3 to eliminate vulnerabilities exposed by attacks such as KRACK. By fully comprehending the underlying mechanisms of these protocols, professionals can make informed decisions regarding security policies and implementations. Moreover, it facilitates the detection and mitigation of potential security breaches, ultimately safeguarding sensitive data and maintaining network integrity.

Learning from the vulnerabilities exploited in WPA and WPA2—like KRACK—is essential for implementing effective security measures. It is imperative to adopt a proactive security mindset and regularly update firmware and software to address known vulnerabilities. This approach, coupled with a solid understanding of the protocols' strengths and weaknesses, allows administrators to minimize risks. This understanding extends beyond simply choosing the right protocol; it necessitates continuous monitoring of network traffic, regular security audits, and the implementation of robust intrusion detection and prevention systems. Proactive measures are significantly more effective and cost-efficient in the long run than responding to already compromised systems. Furthermore, it protects not only the network's infrastructure but also the data of its users. A secure network, ultimately, is a network built on both the understanding of past vulnerabilities and a commitment to future resilience.

The legacy of WPA and WPA2 extends beyond their direct application. They represent pivotal stages in the evolution of Wi-Fi security. The lessons learned from their vulnerabilities—and their eventual replacement by WPA3—have fundamentally shaped the design and implementation of modern wireless security protocols. Understanding these protocols gives network administrators a strong foundation for understanding the principles behind modern security, including the importance of strong encryption, robust key management, and the continual vigilance needed to maintain network security. The evolution of wireless security is a continuous process, and the knowledge gained from studying WPA and WPA2 is crucial for navigating this constantly changing landscape. It isn't enough to merely deploy the latest technologies; a solid understanding of the underlying principles ensures that the network remains secure even as the threat landscape continues to evolve.

2.2 WPA3 Enhancements

WPA3 represents a significant leap forward in Wi-Fi security, addressing critical vulnerabilities present in its predecessors, WPA and WPA2. Unlike its forerunners which relied on the vulnerable TKIP and even the susceptible AES-CCMP encryption, WPA3 employs the more robust SAE (Simultaneous Authentication of Equals) protocol for key exchange. This eliminates the pre-shared key (PSK) vulnerabilities exploited in many past attacks, significantly enhancing the overall security posture of the network. The transition from a four-way handshake to a simultaneous process is a pivotal change, offering enhanced protection against dictionary attacks and brute-force attempts, common methods used to compromise network security. The implementation of SAE is a major departure from older methods, making WPA3 intrinsically more secure. Its strength lies in its resistance to offline dictionary attacks; a hacker intercepting the handshake would find it computationally infeasible to crack the password, thus protecting users from common password-guessing methods. This crucial advancement improves the overall security of user credentials and, by extension, the network's data integrity. Furthermore, the robust encryption algorithms employed by WPA3 provide a high level of confidentiality.

Another notable enhancement in WPA3 is its improved protection against rogue access points. These malicious access points mimic legitimate networks, deceiving users into connecting and allowing attackers to intercept sensitive data. WPA3 mitigates this threat through enhanced authentication protocols and improved network discovery mechanisms. It offers more rigorous checks to verify the authenticity of the access point before establishing a connection, ensuring users connect only to legitimate networks. This greatly reduces the risk of users inadvertently falling prey to man-in-the-middle attacks, a common exploitation technique used in conjunction with rogue access points.

Moreover, WPA3 introduces enhanced security features for public Wi-Fi networks. In the past, public hotspots often lacked robust security measures, making them vulnerable to various attacks. WPA3 addresses this by providing more secure authentication options and improved encryption. The enhanced security features reduce the risks associated with using public

Wi-Fi, allowing users to access the internet with increased confidence. This is crucial given the increasing reliance on public Wi-Fi for both personal and professional use. It builds trust in public Wi-Fi access points, encouraging greater usage while maintaining secure connections.

Beyond enhanced security features, WPA3 introduces improvements in network management. It offers better diagnostics and troubleshooting tools, enabling network administrators to efficiently identify and resolve issues. This aspect contributes to improved overall network uptime and better security management. The improved monitoring allows network administrators to anticipate potential threats and implement preventative measures effectively, leading to more robust network security. This preventative approach to cybersecurity is crucial for maintaining a secure and reliable wireless network infrastructure.

Implementing WPA3 involves several steps. Firstly, ensure your wireless access point supports the WPA3 protocol. Many modern access points are already compatible, but it's vital to verify this through your router's documentation or settings. Upgrading the firmware of older devices may also enable WPA3 support, a crucial step in enhancing network security. Next, configure your access point to use WPA3. The exact process may vary slightly depending on the vendor and access point model. However, the basic steps typically involve enabling WPA3 within the security settings and selecting a strong password.

Selecting a strong password is crucial for maintaining the integrity of the WPA3 security. This aspect should not be overlooked; passwords should include a mix of uppercase and lowercase letters, numbers, and special characters. Using a password manager can help in creating and managing strong, unique passwords for various devices and accounts, a critical step in bolstering overall cybersecurity. Finally, regularly update your access point's firmware to address any identified vulnerabilities, as security updates often include patches for newly discovered vulnerabilities. Keeping the firmware up-to-date is an essential part of maintaining robust network security. Through these simple steps, you can ensure your network is protected by the enhanced security features offered by WPA3. This commitment to proactive security measures will significantly reduce the risk of compromise.

2.3 Open Authentication Risks

Open authentication, while seemingly convenient, presents significant security risks for Wireless Access Points (WAPs). It essentially broadcasts an invitation to any device within range, bypassing the crucial step of user authentication. This open door policy renders the network susceptible to various attacks, potentially compromising sensitive data and disrupting network operations. Imagine a public Wi-Fi hotspot at a cafe – with open authentication, anyone can connect, potentially including malicious actors seeking to intercept data or launch attacks.

The lack of a password or any form of verification mechanism is the primary vulnerability. Attackers can easily connect to the network without authorization, gaining access to resources and potentially compromising user devices. This lack of control extends to the potential for unauthorized network access, allowing malicious users to conduct man-in-the-middle attacks, intercepting communications and stealing sensitive information, such as login credentials, financial data, and personal communications. Consider the implications for businesses with sensitive data transmitted wirelessly; open authentication could expose them to significant financial and reputational damage.

Furthermore, open authentication makes the network vulnerable to denial-of-service (DoS) attacks. A single attacker, or even a botnet, can flood the network with traffic, overwhelming its resources and preventing legitimate users from connecting. This not only disrupts services but can also leave the network exposed to further attacks while resources are depleted attempting to handle the illegitimate traffic. This disruption can significantly impact productivity in a business setting or cause frustration for users in a public setting, leading to lost revenue or damaged reputation.

Beyond DoS attacks, open authentication invites various forms of malware. Malicious actors could create and distribute rogue access points mimicking legitimate networks. Unsuspecting users connecting to these fake networks would unknowingly expose their devices to malware, potentially allowing hackers to steal data or control their devices. This underscores the need for robust security measures beyond simple password protection, especially in high-risk environments.

Even if no immediate malicious activity occurs, the lack of authentication presents an ongoing security risk. Open networks allow for unauthorized access to sensitive internal resources. This can result in data breaches that go undetected for extended periods, making remediation difficult and costly. Data loss is just one consequence; regulatory penalties for non-compliance with data protection regulations could also add a significant financial burden.

Therefore, deploying open authentication on any WAP should be strongly discouraged, particularly in environments where data security and network integrity are paramount. The perceived convenience of open authentication is drastically outweighed by the extensive security risks it presents. Implementing robust authentication mechanisms, such as WPA2 or WPA3, is essential to protect sensitive data, maintain network integrity, and comply with data protection regulations. The cost of a security breach, both financially and reputationally, far outweighs the minimal convenience of open authentication.

Prioritizing security over convenience is crucial. The consequences of a security breach related to open authentication can range from data loss and financial penalties to reputational damage and legal action. Proper security protocols, including strong passwords and robust authentication mechanisms, must be implemented to ensure the safety and integrity of the wireless network and its users. Think of your WAP as a front door to your network; leaving it unlocked (open authentication) is simply inviting trouble. Instead, adopt stringent security measures to safeguard your valuable assets.

3 Network Segmentation Strategies

Network segmentation is crucial for securing your WAPs. Think of it like creating separate, fortified zones within your network. Each zone, or segment, houses specific devices and applications, limiting the impact of a breach. If one segment falls, the others remain protected. This is fundamentally different from a flat network where a single compromised device can offer access to the entire system. This isolation is the cornerstone of robust WAP security.

Implementing VLANs (Virtual Local Area Networks) is your first step. VLANs logically divide your physical network into multiple broadcast domains. You can group devices based on their function – for example, separating guest Wi-Fi from your internal network. This prevents unauthorized access between segments, even if a malicious actor compromises one. Configuration involves assigning specific ports or devices to particular VLANs using your network switch. Careful planning is key to maximizing the effectiveness of your segmentation strategy. Understand your network topology and device roles before implementation, to optimize VLAN assignments for best results. Consider using a network management system for easier oversight and control. Remember, proper VLAN configuration is more than just creating them—it's about effectively segmenting based on risk profiles. Firewalls are your next line of defense, acting as guardians between these VLANs. They meticulously inspect network traffic, allowing only authorized communication between segments. This prevents lateral movement – the ability of an attacker to move from a compromised device to other parts of your network. Configure your firewall rules to strictly restrict access between VLANs. Only allow essential traffic, such as communication between servers and clients that need to interact. Regularly review and update firewall rules to reflect changes in your network infrastructure or security requirements. Ensure robust logging and monitoring capabilities to detect and respond to suspicious activity. Implement strict access control lists to limit what users or devices can access on the network, enhancing security, and enabling better control of traffic flow.

Beyond basic firewall rules, consider advanced features like intrusion prevention systems (IPS) integrated into your firewalls. These actively block malicious traffic, preventing attacks before they can cause damage. Moreover, implementing application control can restrict access to specific applications based on user or device type. This adds another layer of defense against threats targeting vulnerable applications. For instance, blocking certain applications on your guest network prevents unintended access to sensitive data or internal systems. This can minimize the risk to your internal network if the guest network is compromised.

Moving beyond VLANs and firewalls, consider implementing access control lists (ACLs) on routers. ACLs provide granular control over network access, allowing you to define specific rules for which traffic is permitted to pass between different network segments, enabling more specific control. For example, you could create rules that only allow specific IP addresses or ports to communicate between the network's internal and external portions. By carefully implementing and maintaining ACLs, you can significantly enhance the network's security posture by regulating network traffic flow and restricting unauthorized access.

Remember that network segmentation isn't a one-time task; it's an ongoing process. Regularly review your segmentation strategy, ensuring it aligns with your evolving network needs and security posture. New devices, applications, and threats necessitate adjustments to maintain optimal protection. Conduct periodic security audits to identify vulnerabilities and make necessary modifications. Your network is a dynamic entity, and so should your segmentation strategy be. By consistently adapting your segmentation strategy, you can proactively mitigate risks and maintain a robust security posture. This adaptable approach ensures that your network security remains effective against the ever-evolving landscape of cyber threats.

3.1 VLAN Implementation

VLANs offer a powerful method for segmenting your network, significantly enhancing WAP security. Implementing them correctly involves understanding their functionality and carefully planning your network topology. This isn't just about creating separate broadcast domains; it's about strategically isolating sensitive data and critical resources. Consider, for example, separating guest Wi-Fi from your internal corporate network. This simple step drastically reduces the potential impact of a compromise on the guest network. Failure to properly segment could lead to a single breach compromising your entire infrastructure.

Proper VLAN configuration requires careful consideration of your network's structure and the devices involved. You'll need to configure your switches to support VLANs, assigning ports to specific VLANs based on their intended use. This might involve using tagging mechanisms like 802.1Q, which adds VLAN information to the Ethernet frame header. Careful planning is crucial here; a poorly planned VLAN setup can create more problems than it solves. You must meticulously document your VLAN assignments.

Beyond simple port assignments, VLANs can be integrated with other security mechanisms. Firewalls, for instance, can be configured to control traffic flow between VLANs, acting as a further layer of protection. This allows you to restrict communication between specific VLANs, preventing unauthorized access and the lateral movement of attackers. For instance, you could restrict access from the guest VLAN to your server VLAN, limiting potential damage from a successful compromise. Remember, security is often layered – VLANs alone are not a complete solution.

One common pitfall is assigning too many devices to a single VLAN, negating the benefits of segmentation. Overcrowded VLANs become just as vulnerable as a single, unsegmented network. Aim for logically grouped devices, minimizing the impact of a potential breach within a specific VLAN. Consider the potential impact of each device's compromise—a compromised workstation in a single VLAN is less damaging than one with access to your database server.

The management of VLANs requires ongoing maintenance. Periodic reviews of your VLAN configuration ensure it aligns with your evolving network needs and security posture. Regular audits can identify redundant or inefficient VLANs, optimizing performance and security. Changes to your network, such as adding new devices or departments, should trigger a re-evaluation of your VLAN structure. This is not a "set and forget" process; proactive management is crucial.

Consider also the impact of mobility on VLAN assignments. If your network uses roaming clients, such as laptops, you need a mechanism for dynamically assigning VLANs based on the client's location or role. This dynamic assignment adds complexity, requiring careful consideration of your wireless infrastructure and authentication methods. Remember, the goal is seamless access with consistent security. Poorly implemented dynamic VLAN assignment can lead to security gaps.

Finally, remember the importance of thorough testing after any VLAN configuration changes. This involves checking connectivity between VLANs, ensuring proper isolation and controlled access. You should rigorously test the security measures associated with your VLANs to confirm that traffic is appropriately controlled and unauthorized access is effectively prevented. Testing includes simulating attacks to verify the effectiveness of your security controls in preventing unauthorized access or movement between VLANs. A well-tested VLAN configuration gives you peace of mind. A poorly tested configuration invites unnecessary risk.

3.2 Firewall Configuration

Firewalls are essential for securing your wireless access point (WAP). They act as gatekeepers, controlling network traffic based on predefined rules. Think of them as highly selective bouncers, allowing only authorized individuals and data to enter your network's VIP section. Improperly configured firewalls, however, can leave gaping holes in your security, rendering them essentially useless. Effective firewall configuration requires meticulous attention to detail and a solid understanding of your network's needs.

A crucial first step is defining your network's zones. This involves segmenting your network into logical units, like a guest network, a corporate network, or a dedicated server network. Each zone should have its own distinct security policies, controlled by the firewall. For instance, the guest network might have limited access to internal resources, while the corporate network requires stricter controls. Careful planning of these zones reduces the impact of a successful breach, confining attackers to a specific section rather than granting them free reign across your entire network. You'll want to visually represent this segmentation; a network diagram proves invaluable in this planning process. Consider including devices, subnets, and the relationship between them.

Next, you need to establish robust access control lists (ACLs). ACLs are the rules that dictate which traffic is permitted or denied. They specify the source and destination IP addresses, ports, and protocols, allowing granular control over network access. For example, you might allow web traffic from the internet to your web server but block all other inbound connections. Simultaneously, you might allow internal users to access all necessary resources, while restricting access for guest users. Carefully consider the specifics of each rule; an improperly configured ACL can easily negate the entire security posture of your firewall. Don't be tempted to apply overly broad or permissive rules to simplify the configuration process; this approach significantly compromises security.

Don't underestimate the importance of regular updates and maintenance. Firewall vendors regularly release updates that patch security vulnerabilities and improve performance. Failing to update your firewall leaves it susceptible to known exploits, negating any security gains from careful configuration. Furthermore, regular monitoring of firewall logs is crucial. These logs provide a detailed record of all network traffic that interacts with the firewall. Regular review can alert you to suspicious activity, potential breaches, and identify misconfigurations that might need immediate attention. Automated alerts for critical events can significantly enhance this monitoring process, providing immediate notification for any anomalies.

Stateful inspection is a powerful technique used by modern firewalls to enhance security. It tracks the state of network connections, allowing only traffic that is part of an established connection. This prevents many types of attacks that rely on unsolicited inbound connections. By verifying that incoming traffic is a legitimate response to a prior outbound connection, stateful firewalls greatly reduce the likelihood of successful infiltration. Implementing stateful inspection as a key element in your firewall ruleset is essential for mitigating many common vulnerabilities.

Consider employing intrusion prevention systems (IPS). IPSs work in conjunction with firewalls, actively detecting and blocking malicious traffic based on signatures of known attacks or unusual patterns. An IPS enhances the capabilities of a firewall by adding a proactive layer of defense against sophisticated attacks that may bypass traditional firewall rules. IPS frequently updates their threat databases; integrating this function into your routine maintenance enhances the effectiveness of your firewall setup.

Finally, remember that even the best firewall configuration is not foolproof. It's a critical component of a layered security approach, working in tandem with other security measures such as intrusion detection systems, strong passwords, and regular security audits. No single security measure guarantees complete protection; instead, rely on a combination of strategies to mitigate risks and improve your overall security posture. A comprehensive security strategy ensures that even if one layer of defense fails, other layers remain to protect your network. Regularly review and refine your approach; the security landscape is constantly evolving.

4 Advanced Encryption Techniques

AES, or Advanced Encryption Standard, forms the bedrock of many modern encryption systems. Its strength lies in its symmetric key structure, meaning the same key is used for both encryption and decryption. Different key sizes—128, 192, and 256 bits—offer varying levels of security, with 256-bit keys currently considered virtually unbreakable with current technology. Understanding these variations is crucial for choosing the appropriate level of protection based on the sensitivity of the data being transmitted. For example, a high-security government network would demand the strongest 256-bit encryption, while a less sensitive home network might suffice with 128-bit. Remember, the longer the key, the more computationally intensive it is to crack.

Effective key management is paramount; without it, even the strongest encryption algorithm is vulnerable. Secure key generation, storage, and distribution are vital. Consider using hardware security modules (HSMs) for storing sensitive cryptographic keys, as they offer significantly enhanced protection against physical theft or unauthorized access. Regular key rotation is also essential. Changing keys frequently reduces the window of vulnerability, limiting the potential damage if a key is compromised. Implementing strong access controls to key management systems is vital, restricting access to only authorized personnel. These robust processes are non-negotiable for maintaining a high level of security.

Public Key Infrastructure (PKI) offers a solution for secure key exchange in a networked environment. Unlike symmetric encryption, PKI utilizes a pair of keys: a public key for encryption and a private key for decryption. The public key can be widely distributed, allowing anyone to encrypt data destined for the owner of the corresponding private key. Only the holder of the private key can decrypt the message, ensuring confidentiality. PKI's implementation hinges on trusted certificate authorities (CAs) that vouch for the authenticity of public keys. This trust framework is essential for secure online transactions and communication. It's crucial to use a reputable CA and to ensure the validity of certificates before establishing trust.

Perfect Forward Secrecy (PFS) adds another layer of security, ensuring that the compromise of a long-term key doesn't compromise past communications. With PFS, each session uses a unique, ephemeral key, even if the long-term key is used for initial authentication. If the long-term key is compromised, previous sessions remain protected because their unique keys are unrelated. Implementing PFS is critical for securing sensitive communications, particularly in environments where long-term keys might be vulnerable to attacks. Its effectiveness relies on robust key exchange protocols and secure key generation mechanisms. Consider algorithms such as Diffie-Hellman or Elliptic Curve Diffie-Hellman (ECDH) for implementing PFS.

Elliptic Curve Cryptography (ECC) offers comparable security to RSA with smaller key sizes, making it more efficient for resource-constrained devices like mobile phones and embedded systems commonly found in IoT devices. This efficiency is crucial in wireless networks where bandwidth and processing power might be limited. ECC relies on the mathematical properties of elliptic curves to create strong cryptographic algorithms for digital signatures and encryption. Its smaller key sizes translate to faster encryption and decryption speeds and reduced storage requirements, all vital considerations in the demanding world of wireless security. When choosing an encryption algorithm, remember to consider the efficiency alongside the security requirements of your particular WAP deployment. The efficiency and security trade-off is an important consideration when deciding whether to use ECC over traditional methods like RSA.

4.1 AES & Its Variations

The Advanced Encryption Standard (AES), a symmetric-key encryption algorithm, forms the backbone of many modern security systems, including those protecting Wireless Access Points (WAPs). Its strength lies in its efficiency and robustness against known cryptanalytic attacks. Understanding AES is crucial for anyone involved in WAP security. At its core, AES operates on a block cipher model, encrypting data in 128-bit blocks. The key size, however, can vary, offering 128-bit, 192-bit, and 256-bit options; each offering progressively stronger security against brute-force attacks. The longer the key, the more computationally intensive it is to crack.

AES, in its various key lengths, isn't simply a single algorithm but a family of algorithms. This family shares the same fundamental design principles but differs in the number of rounds used in the encryption and decryption processes. These rounds involve several transformations, including substitution, permutation, and key mixing. The number of rounds increases with the key size, ensuring that even with larger keys, the data remains securely encrypted, withstanding advanced attacks for a longer period. This scalability allows it to adapt to different security needs, making it versatile for numerous applications.

AES's popularity stems from its public vetting process, a rigorous evaluation conducted by independent experts before its adoption by the National Institute of Standards and Technology (NIST). This transparency and thorough testing instilled confidence in its security, leading to its widespread adoption. This public scrutiny and rigorous testing distinguish AES from other encryption algorithms which may not have undergone such comprehensive review. Consequently, AES remains the gold standard for symmetric-key encryption, constantly being improved.

However, the implementation of AES is only as strong as its key management practices. Secure key generation, storage, and distribution are paramount. Weak keys, carelessly managed, can render even the strongest algorithm vulnerable. For instance, a poorly generated key can be easily guessed, negating AES's strength entirely. Therefore, robust key management practices are absolutely necessary for effective data protection. This is often where vulnerabilities are exploited, regardless of the robustness of the encryption algorithm itself.

AES variations aren't merely about different key sizes. They extend to specific modes of operation, which dictate how AES handles data beyond the 128-bit block size. These modes address different security and performance requirements. For example, Cipher Block Chaining (CBC) mode chains the previous ciphertext block into the encryption of the current block, adding a level of diffusion. But CBC mode suffers from a vulnerability known as padding oracle attacks, highlighting the importance of understanding the nuances of each mode's security. Furthermore, other modes, like Counter (CTR) mode, offer advantages in terms of parallelization and non-blocking operations, making them suitable for high-performance applications.

Understanding the strengths and weaknesses of different AES modes is critical for selecting the optimal configuration for a given WAP security scenario. For instance, CTR mode might be preferred for high-throughput applications, while GCM (Galois/Counter Mode) mode, which provides authenticated encryption, is a more suitable choice where both confidentiality and data integrity are necessary. The selection of the appropriate mode depends on factors such as performance requirements, security level needed, and the overall system architecture. Therefore, simply choosing AES is insufficient; a deep understanding of its operational modes is essential for robust security implementation.

While AES is exceptionally secure when implemented correctly, it's not impervious to future threats, particularly those posed by quantum computing. This has spurred research into post-quantum cryptography, algorithms designed to withstand attacks from quantum computers. Although AES remains robust against classical attacks, planning for the long-term requires staying informed about advancements in cryptography and considering strategies for migrating to quantum-resistant algorithms when they become widely available and accepted as suitable replacements. Therefore, adopting a forward-thinking approach and staying up-to-date with emerging research is crucial for maintaining long-term WAP security.

In conclusion, the widespread adoption of AES demonstrates its importance in modern network security, particularly within the context of WAP protection. However, mastering AES involves more than simply understanding its underlying mathematical principles. It also requires a comprehensive grasp of its variations, modes of operation, and the crucial aspects of key management. Failing to address these facets can severely compromise the effectiveness of even the strongest encryption algorithm. A deep understanding of AES, including its variations, is not just about understanding the technology itself, but understanding its practical implications and potential vulnerabilities within the wider context of WAP security. Continuously educating oneself on advancements in this field is essential to ensure the continued security of wireless networks.

4.2 Key Management Practices

Effective key management is paramount to the security of any wireless access point (WAP). Without robust key management practices, even the strongest encryption algorithms are vulnerable. Think of it like a highly fortified castle with a single, easily accessible gate – the encryption is the castle walls, but the key management is the gate. A compromised key renders the entire system insecure, irrespective of the encryption's strength.

Implementing a strong key management system requires a multi-faceted approach. This isn't a single solution, but rather a carefully orchestrated combination of strategies, demanding careful consideration and meticulous execution. Neglecting any one aspect can severely weaken the overall security posture. We must strive for layers of protection, creating a defense in depth. Regular key rotation is fundamental. This involves changing encryption keys at predetermined intervals. The frequency depends on the sensitivity of the data being transmitted and the potential risk level. For highly sensitive data, more frequent rotations are crucial, perhaps even daily. Less sensitive networks might allow for weekly or monthly changes, but consistency is key. Failing to establish a rigorous key rotation schedule leaves your system exposed to protracted attacks. A well-defined process is critical, ensuring all devices obtain the updated keys seamlessly, preventing disruptions in service.

Secure key storage is equally critical. Keys should be stored in a protected, tamper-proof environment, accessible only to authorized personnel. Using hardware security modules (HSMs) is an excellent practice, offering robust physical and cryptographic protection. These dedicated hardware devices are specifically designed to securely manage and process cryptographic keys, significantly reducing the risk of compromise. Simply storing keys on a regular server is inadequate; it leaves them vulnerable to malware and unauthorized access. Employing strong access control mechanisms, including multi-factor authentication, is an absolute necessity.

Key distribution procedures require careful planning. Using secure methods like out-of-band key exchange, or employing key distribution centers (KDCs) within a trusted network infrastructure, minimizes the chance of interception during transmission. Employing insecure channels for key distribution is akin to sending the castle keys through the mail – exceptionally risky. Remember, the process isn't about just choosing strong keys; it's about securely managing their entire lifecycle, from creation to destruction. Auditing key usage and monitoring for any irregularities are important aspects of a comprehensive key management strategy. Regular audits provide insights into potential vulnerabilities or compromises. Anomaly detection systems can flag suspicious activity, enabling rapid responses. For example, a sudden increase in failed login attempts using a specific key might signal a breach attempt, allowing for immediate key revocation and mitigation of any further compromise.

Furthermore, proper key revocation mechanisms are essential. When a key is suspected of being compromised, a swift and effective revocation process must be in place. This involves immediately disabling the compromised key, preventing further unauthorized access. Effective revocation necessitates a robust system for disseminating the information about the revoked key to all relevant systems and devices. The process must be simple, efficient, and reliable, minimizing any downtime.

Finally, remember the human factor. Employee training on secure key handling practices is paramount. Even the most robust technical measures can be undermined by human error. Regular training sessions and awareness programs should educate employees on the importance of key security, proper password management, and recognizing phishing attempts. The overall success of key management relies on a combination of technology and human diligence; one without the other creates a significant weakness.

Therefore, a well-defined key management strategy demands ongoing attention, continuous improvement, and regular audits to ensure your WAP remains secure.

4.3 Public Key Infrastructure

Public Key Infrastructure (PKI) is a cornerstone of secure communication in many contexts, including securing wireless access points (WAPs). It provides a framework for managing digital certificates, which act as electronic identities for devices and users. Think of it as a digital passport system for your network, enabling secure authentication and encrypted communication without needing to exchange secret keys directly, a significant advantage in the often-open world of wireless networking. This system offers several benefits, making it crucial for robust WAP security.

Unlike symmetric encryption, which relies on the same secret key for encryption and decryption, PKI uses a pair of keys: a public key, freely distributed, and a private key, kept strictly confidential. This asymmetrical approach allows anyone to encrypt data using the public key, but only the holder of the corresponding private key can decrypt it. This elegant solution resolves the key distribution problem inherent in symmetric systems, a major vulnerability when securing numerous devices within a wireless network.

The core components of PKI are Certificate Authority (CA), Registration Authority (RA), and certificate revocation lists (CRLs). The CA issues and manages digital certificates, acting as the trusted root of the entire PKI structure. The RA verifies user identities before the CA issues certificates, ensuring only legitimate entities receive them. CRLs track revoked or compromised certificates, preventing their use in authentication or encryption. This robust three-layered architecture, including a constant monitoring process, ensures the integrity of your WAP security.

Implementing PKI in a WAP environment involves several practical steps. First, you must choose a reputable CA to issue certificates for your WAP and all the devices that need to access it. Next, each device requires a unique digital certificate, properly configured and installed on that device. This involves generating key pairs, submitting a certificate signing request (CSR) to the CA, and installing the issued certificate. Finally, you'll need to configure your WAP to trust the CA's certificate and to use digital certificates for authentication and encryption. This detailed process strengthens the security of your wireless network considerably.

A crucial aspect of PKI is certificate lifecycle management. Certificates have expiration dates, and they might need to be renewed regularly to maintain security. Furthermore, if a private key is compromised, the corresponding certificate must be revoked immediately. Monitoring these certificates is essential; this is commonly done using CRLs, but some use Online Certificate Status Protocol (OCSP) for real-time verification. Failing to manage the certificate lifecycle effectively will severely weaken your security posture, negating many of the advantages of a PKI setup.

Furthermore, PKI is fundamental for securing other aspects of your WAP security strategy, such as secure management access and the use of virtual private networks (VPNs). A secure management protocol like HTTPS, enabled by PKI, safeguards your WAP's configuration from unauthorized

changes. Similarly, implementing VPNs often relies on PKI for secure authentication and encryption of the tunnel traffic. Therefore, having a robust PKI implementation complements your overall wireless network security, providing a much more complete solution.
Remember, the strength of your PKI implementation hinges on the security of your private keys. Protect them zealously. Use strong key generation practices, ensure the safety of your key storage mechanisms, and follow best practices on user management. Incorporating strong authentication measures, multi-factor authentication, if possible, is essential to improve this security even further. Regularly review and update your PKI setup and associated configurations to counter emerging threats and to keep your system current. A well-managed PKI system drastically reduces risks, contributing to a comprehensive WAP security architecture. Implementing PKI adds a significant layer of protection, significantly improving your wireless network's security.

4.4 Perfect Forward Secrecy

Perfect Forward Secrecy, or PFS, is a crucial element in securing communications, particularly within the context of Wireless Access Points (WAPs). Its primary function is to ensure that the compromise of a long-term key doesn't retroactively compromise past sessions. Imagine a scenario where an attacker gains access to a server's private key; with PFS in place, they wouldn't be able to decrypt past conversations, even with this significant breach. This forward-looking protection is vital for maintaining confidentiality and integrity.

Understanding PFS requires grasping its fundamental mechanism. Unlike systems that reuse the same session key for extended periods, PFS leverages ephemeral keys for each session. These temporary keys are generated uniquely for every connection, and they're discarded once the session ends. This dynamic approach ensures that even if a long-term key is exposed, the attacker is left with nothing more than a single session key, rendering the decryption of past communications effectively impossible. This principle forms the bedrock of robust security.

The implementation of PFS hinges on the use of Diffie-Hellman key exchange or its elliptic curve variant, Elliptic Curve Diffie-Hellman (ECDH). These algorithms allow two parties to securely establish a shared secret key over an insecure channel, without ever needing to directly transmit the secret. This secret key, then, is used to encrypt the session, while the long-term keys remain untouched and solely used for authentication. This is why ECDH is considered a keystone in achieving robust PFS. Understanding these key exchange mechanisms is fundamental to understanding PFS.

Consider a practical example: two devices are attempting to communicate securely through a WAP. With PFS enabled, each device generates its own ephemeral key. They then exchange information via the Diffie-Hellman algorithm, resulting in a shared secret. This shared secret becomes the session key, exclusively used for the encryption and decryption of the communication during that specific session. If an attacker later compromises the long-term key used for authentication, they cannot decrypt past conversations because they lack the ephemeral keys used to generate the session keys. It's this ephemeral nature that is absolutely crucial.

Several protocols support PFS, and understanding their capabilities is essential for effective security. Transport Layer Security (TLS) and its predecessor Secure Sockets Layer (SSL) versions 1.2 and higher inherently support PFS, primarily using Diffie-Hellman or ECDH. When configuring your WAP, ensure that these protocols are enabled and configured to use strong ciphers that support PFS. Failing to do so leaves your network vulnerable to historical data breaches, negating a considerable layer of security. Careful protocol selection and configuration are paramount.

Choosing the appropriate key exchange algorithm is also critical. While Diffie-Hellman is well-established, ECDH is generally preferred today due to its improved efficiency and comparable security with smaller key sizes. This efficiency is significant, especially for resource-constrained devices often found in wireless environments. Therefore, when selecting your cipher suite, prioritize those that employ ECDH for its performance advantages without compromising security. Remember, choosing the right cipher suite determines the strength of the system.

Beyond the technical implementation, the importance of PFS lies in its contribution to long-term security posture. In an increasingly interconnected world, where data breaches are unfortunately common, PFS acts as a critical safeguard against future compromises. Think of it as an insurance policy – the expense of implementing it is far outweighed by the potential costs of a breach where past communications are exposed. The peace of mind it affords is also immeasurable.

Finally, it's crucial to emphasize the importance of regular security audits. These audits should explicitly assess the effectiveness of PFS implementation. Outdated or improperly configured systems are vulnerable, rendering PFS ineffective. Staying updated on the latest security best practices and leveraging tools to monitor the strength of the cryptographic keys used is also an essential part of maintaining a strong security posture. Proactive monitoring and timely updates are key to continued protection. Therefore, incorporating PFS and understanding its limits are crucial for modernizing your network security strategy, creating a more secure and trustworthy wireless environment. The investment in both implementation and ongoing maintenance is a vital component of comprehensive WAP security.

4.5 Elliptic Curve Cryptography

Elliptic curve cryptography (ECC) offers a compelling alternative to traditional public-key cryptosystems like RSA, particularly relevant in resource-constrained environments like many WAP implementations. Its strength lies in its ability to achieve comparable security levels with significantly smaller key sizes, leading to faster computation and reduced bandwidth consumption—factors crucial for efficient wireless communication. This efficiency stems from the mathematical foundation of ECC, which utilizes the properties of elliptic curves over finite fields. These curves, defined by specific equations, create a group structure where points can be added and multiplied, forming the basis for cryptographic operations.

The core of ECC involves generating a pair of keys: a private key and a public key. The private key is a randomly selected integer, kept secret, while the public key is derived from the private key through multiplication on the elliptic curve. This asymmetric nature means that knowing the public key doesn't reveal the private key, a critical element for secure communication. Encryption and decryption processes involve point multiplication and addition on the elliptic curve, leveraging the inherent difficulty of the discrete logarithm problem—finding the private key given the public key. This mathematical challenge underpins the security of ECC, making it resistant to attacks.

Compared to RSA, which relies on the difficulty of factoring large numbers, ECC offers a more compact solution. A 256-bit ECC key provides comparable security to a 3072-bit RSA key. This substantial size difference translates directly into performance advantages, especially valuable in mobile devices and other devices with limited processing power. The smaller key sizes mean faster key generation, encryption, and decryption processes, minimizing latency and improving overall performance in a WAP context, where timely responses are critical.

Implementing ECC in WAP security involves several practical steps. First, select an appropriate elliptic curve and associated parameters. Standardized curves, like those specified by NIST, provide a solid foundation, ensuring interoperability and leveraging the extensive scrutiny these curves have undergone. Next, generate key pairs, securely storing the private key and distributing the public key appropriately. This process demands robust key management practices, including secure key generation, storage, and distribution protocols, to prevent compromise of the private key.

The integration of ECC into WAP security protocols is typically done at the application layer, often employing established protocols such as TLS/SSL. These protocols leverage ECC for key exchange, ensuring secure communication between devices. The handshake process involves the exchange of public keys, allowing both sides to establish a shared secret key used for symmetric encryption of the data. Moreover, the use of ECC can enhance the security of existing authentication mechanisms, providing more robust authentication and authorization procedures.

However, implementing ECC isn't without its challenges. Proper implementation requires careful consideration of various factors. Selecting secure parameters, utilizing robust key management techniques, and ensuring correct implementation are crucial for preventing vulnerabilities. Errors in implementation can lead to weaknesses exploitable by attackers. It's vital to use well-vetted cryptographic libraries and thoroughly test the implementation for security flaws. Furthermore, regular updates and patches are needed to address any newly discovered vulnerabilities in the underlying cryptographic algorithms or the implementation itself.

Finally, ongoing research and development in the field of ECC contribute to its continuous improvement and strengthening. The cryptographic community actively monitors for any weaknesses, ensuring the continued security of this vital encryption technique. Staying informed about the latest developments and adopting best practices are crucial to maintaining the highest level of security. As wireless technology continues to advance, ECC will undoubtedly remain a pivotal cryptographic technique, providing robust security for sensitive data transmitted over WAPs. Its efficiency and security characteristics make it an ideal choice for securing the wireless landscape, guaranteeing secure communication in an increasingly connected world.

5 Intrusion Detection & Prevention

Deploying an Intrusion Detection/Prevention System (IDS/IPS) is crucial. Think of an IDS as a vigilant security guard, passively monitoring network traffic for suspicious activity, alerting you to potential threats. An IPS goes further; it actively blocks or mitigates these threats, acting as a proactive defender. Choosing between the two depends on your specific needs and resources; an IPS offers stronger protection, but an IDS can be a more cost-effective starting point for smaller networks. Proper placement within your network is paramount; consider placing them at strategic points, such as the gateway to your wireless network, to effectively monitor all incoming and outgoing traffic. Remember that a well-configured firewall remains a critical first line of defense; the IDS/IPS complements, not replaces it.

Wireless anomaly detection plays a vital role in identifying unusual patterns indicating potential intrusions. These systems leverage statistical analysis and machine learning algorithms to establish a baseline of normal network behavior. Any deviation from this baseline—unexpected traffic spikes, unusual device connections, or atypical data flows—triggers an alert, enabling prompt investigation and response. Implementation requires careful configuration and fine-tuning; you need to define what constitutes an "anomaly" specific to your network's typical activity. False positives are a common challenge; regular review and adjustments are essential to optimize performance and minimize disruption caused by legitimate but unusual network events. Integrating anomaly detection with your IDS/IPS enhances overall effectiveness, providing layered protection.

Security Information and Event Management (SIEM) systems are your centralized command center for security monitoring. They collect and correlate security logs from various sources, including your IDS/IPS, firewalls, and other network devices. This consolidated view provides a comprehensive picture of your network's security posture, allowing for faster threat detection and response. The ability to analyze this data reveals trends and patterns that might otherwise go unnoticed, facilitating proactive security measures. Effective SIEM implementation requires careful planning; data volume can be substantial, necessitating appropriate storage and processing capabilities. Furthermore, configuring alerts and dashboards to highlight critical events is crucial to ensure timely responses to potential threats. Successful SIEM usage requires skilled personnel capable of analyzing the complex data and translating it into actionable insights. It provides invaluable context for security incidents, allowing for more informed decision-making and improved incident response times. Ultimately, a well-integrated SIEM strengthens your overall security posture significantly.

5.1 IDS/IPS Deployment

Deploying an Intrusion Detection/Prevention System (IDS/IPS) within a Wireless Access Point (WAP) security architecture requires careful planning and execution. Your initial step involves identifying the specific security needs of your network. Consider factors such as the number of users, the sensitivity of the data transmitted, and the level of threat you anticipate. This assessment will guide your choice between a Network-based IDS/IPS, which monitors network traffic for suspicious patterns, or a Host-based IDS/IPS, which monitors individual devices for malicious activities. A well-defined strategy allows you to optimize your system for maximum efficiency and minimal resource consumption. Understanding your network's traffic patterns and potential threats is crucial for effective IDS/IPS deployment.

Accurate placement of your IDS/IPS is critical for optimal performance. Network-based IDS/IPS devices are typically placed strategically within the network, often at points where traffic converges, such as at the gateway or at crucial junctions within your VLAN structure. This ensures that as much traffic as possible is monitored. Host-based IDS/IPS deployment focuses on individual devices. Consider deploying host-based IDS/IPS on servers, workstations, or other critical devices that hold sensitive information, providing an added layer of protection against attacks targeting those systems specifically. This layered approach combines the broad view of the network-based system with the fine-grained detail of host-based monitoring.

After the physical installation, configuration is just as crucial. Configure your IDS/IPS to monitor the appropriate protocols and ports. For example, if your network primarily uses HTTP and HTTPS, ensure the IDS/IPS monitors traffic on ports 80 and 443. This prevents your system from being overloaded with irrelevant data while still providing adequate coverage for essential traffic. Furthermore, carefully tune the sensitivity settings of your IDS/IPS to balance detection accuracy with the minimization of false positives. Too much sensitivity can lead to an overwhelming number of false alarms, hindering your ability to respond to real threats; too little sensitivity can cause legitimate attacks to be missed. Finding this balance is a key skill for successful IDS/IPS management.

Effective IDS/IPS management includes regular maintenance and updates. The security landscape constantly evolves, with new threats emerging frequently. Therefore, consistently update your IDS/IPS's signature database to ensure it can detect the latest threats. This is paramount for keeping your network protected against the newest attack vectors. In addition, regularly review the logs generated by your IDS/IPS, investigating any suspicious activities flagged by the system. This process helps in the detection of both known and unknown threats. It helps in keeping abreast of potential threats and proactively addressing security vulnerabilities before they can be exploited.

Beyond the technical aspects, integrating your IDS/IPS into your overall security strategy is essential. The data collected by your IDS/IPS provides valuable insights into your network's security posture. You can use this data to identify patterns, understand common vulnerabilities, and adjust your defenses accordingly. This continuous feedback loop allows for adaptive security, crucial for reacting to evolving threats. Furthermore, your

IDS/IPS should be integrated with your Security Information and Event Management (SIEM) system to centralize security monitoring and incident response. This integration enables a unified view of your security environment, facilitating quicker reaction times and more efficient threat remediation. This centralized approach optimizes the effectiveness of your entire security system.

Regular testing and evaluation of your IDS/IPS are integral to its long-term effectiveness. Simulate attacks to evaluate the system's ability to detect and respond to threats accurately. This process identifies any weaknesses in your deployment and configuration, allowing you to address them proactively. Regular testing allows for continuous improvement of your security defenses, and prevents blind spots from undermining your protection strategies. Analyzing your IDS/IPS performance metrics will help you ensure that the system is running efficiently and effectively, providing critical insights into areas for improvement. This proactive and iterative approach ensures your network remains consistently secure and able to withstand potential threats.

5.2 Wireless Anomaly Detection

Wireless anomaly detection is crucial for maintaining the integrity of your wireless network. It involves identifying unusual patterns or behaviors that deviate from established norms, potentially indicating malicious activity or system malfunction. Unlike traditional signature-based intrusion detection systems which rely on known attack patterns, anomaly detection employs statistical and machine learning techniques to discern deviations from the expected operational profile. This proactive approach allows for the detection of zero-day exploits and novel attack vectors that evade signature-based systems, adding a critical layer to your WAP security strategy.

Effective anomaly detection hinges on establishing a robust baseline of normal network behavior. This involves a period of observation to collect data about typical traffic patterns, user activity, and device connectivity. Consider factors like data volume, communication frequency, and device locations. The more comprehensive your baseline, the more sensitive and accurate your anomaly detection system will be. Insufficient data collection can lead to false positives and a diminished ability to distinguish true threats from benign fluctuations. Remember that the initial setup requires careful planning and a period of data gathering to accurately establish a reliable baseline.

Several methods exist for identifying anomalies. Statistical techniques analyze data distributions to find outliers. For instance, a sudden spike in data packets from an unusual source might be flagged. Machine learning algorithms, particularly unsupervised learning methods like clustering and anomaly detection algorithms, can identify patterns not explicitly defined in advance. These algorithms learn from the historical data to identify deviations from normal behavior. Neural networks, for example, can detect subtle anomalies that might be missed by simpler statistical methods, thereby enhancing your network's security posture.

Implementing an anomaly detection system often involves using specialized software or integrating it with your existing network management tools. Some Network Monitoring Tools include basic anomaly detection capabilities; however, for comprehensive analysis, you may need dedicated solutions, which may require specialized expertise for effective implementation and management. Consider the scalability and integration capabilities of any solution before deployment; ensuring compatibility with your current infrastructure is critical. Thorough evaluation of potential solutions is crucial to select a system optimized for your network environment and its specific needs.

Interpreting the results of anomaly detection is critical. A simple alert does not always signify a genuine threat. False positives can arise from legitimate network events mistaken for malicious behavior. Thorough investigation of each alert is crucial. This might involve reviewing network logs, examining device activity, and analyzing the context of the suspected anomaly. A well-trained team will effectively differentiate benign events from security threats. Involving multiple analysts or using automated verification processes can help mitigate the likelihood of false-positive disruptions.

The importance of regular updates and refinements cannot be overstated. As your network evolves and its usage patterns change, your baseline data needs updating to maintain accuracy. Additionally, new attack techniques constantly emerge, making it essential to regularly adapt your detection methods. Regularly review your anomaly detection system's performance, identifying areas where it may be lacking and promptly adapting the model to account for new situations. Keeping your system updated not only helps maintain accuracy but also ensures that your network stays protected against constantly evolving threats.

Beyond the technical aspects, human factors significantly influence the success of anomaly detection. Your team needs the proper training to interpret alerts, investigate incidents, and respond to potential threats effectively. Establish clear incident response protocols, which detail actions to take when an anomaly is detected. Regular training and security awareness programs are important to improve your teams' response time and ability to handle situations quickly and effectively. Regular review of procedures will help your organization's security posture. Remember that the effectiveness of your wireless security depends significantly on human capital and training.

5.3 Security Information & Event Management

Security Information and Event Management (SIEM) systems are crucial for effective WAP security. They provide a centralized platform for collecting, analyzing, and correlating security logs from various network devices, including WAPs, firewalls, and intrusion detection systems. This consolidated view allows security analysts to identify patterns and anomalies indicative of potential threats, enabling quicker response times and minimized downtime. A robust SIEM solution can significantly enhance your overall security posture.

Effective SIEM implementation begins with careful planning and selection of appropriate hardware and software. Consider factors such as the volume of log data generated by your network, the required level of detail in analysis, and the scalability of the chosen system. A system that's too small will become overwhelmed and lead to missed alerts; one that's too large can be unnecessarily expensive and complex. Proper configuration is paramount –

ensuring all necessary devices are correctly connected and forwarding logs to the SIEM platform. The success hinges on meticulous attention to detail during this phase.

Beyond initial setup, ongoing maintenance is vital. Regular updates to the SIEM software are essential to patch vulnerabilities and incorporate improvements. Similarly, you must consistently review and refine the rules and alerts to ensure they accurately reflect evolving threats and your network's specific requirements. Ignoring these updates can render your SIEM ineffective against the latest attack methods. Think of it as a living, breathing component of your defense system, requiring constant care and attention.

Data analysis within a SIEM system should go beyond mere passive monitoring. Security analysts should actively investigate alerts, correlating information from different sources to understand the context and severity of potential threats. This often involves utilizing the system's reporting and visualization capabilities to create customized dashboards and reports that provide insights into trends and patterns in security events. This proactive approach, in contrast to solely reactive measures, allows for earlier identification and mitigation of threats.

The value of a SIEM system extends beyond threat detection. It serves as a valuable tool for compliance audits, providing verifiable evidence of security measures implemented and events that occurred. Being able to show that all appropriate logs have been collected and analyzed is instrumental in demonstrating compliance with relevant regulations and industry standards, reducing the risk of penalties or legal action. This compliance is essential for many businesses.

However, SIEM systems are not a silver bullet. Their effectiveness hinges on several factors, including the quality of the data collected, the expertise of security analysts, and the overall design of the network's security architecture. Relying solely on a SIEM system without implementing other security measures, such as firewalls, intrusion detection systems, and strong access control policies, is a recipe for vulnerability. It is an integral part of a larger strategy.

Therefore, integrating your SIEM system with other security tools enhances its capabilities. For instance, automated responses can be triggered based on specific events or alert thresholds, such as automatically blocking malicious IPs or resetting compromised accounts. This seamless integration is key to leveraging the full potential of your SIEM and optimizing your response time. Consider it a central command center within your broader security ecosystem.

Finally, effective SIEM implementation requires skilled personnel. Security analysts must possess the knowledge and skills to effectively analyze security logs, identify and prioritize threats, and respond accordingly. Investment in training and professional development for your security team is a crucial element of maximizing your SIEM's return on investment. It is not just a software solution; it requires trained individuals to interpret it. This is a significant factor often overlooked.

6 WAP Security Audits & Best Practices

Regular security audits are paramount for maintaining a robust WAP security posture. Think of it as a regular health check for your network; identifying potential problems before they escalate into major security incidents. Vulnerability assessments, the first crucial step, involve systematically scanning your WAPs for known weaknesses. These scans use automated tools and manual analysis to pinpoint vulnerabilities like outdated firmware, weak passwords, or misconfigured settings. Addressing these vulnerabilities promptly is key to preventing exploitation.

Penetration testing takes a more aggressive approach, simulating real-world attacks to expose any exploitable weaknesses that vulnerability assessments might miss. This involves attempting to compromise the WAP using various techniques, mirroring how a malicious actor might proceed. The insights gained from penetration testing are invaluable in strengthening your defenses against sophisticated attacks, uncovering hidden vulnerabilities not detectable through simple scans, providing a realistic picture of your network's resilience. Documenting findings meticulously, along with remediation steps, is crucial for building a comprehensive security report.

Beyond formal audits, consistent monitoring is vital. Regularly review your access logs to identify any suspicious activity, such as unauthorized access attempts or unusual traffic patterns. Implement strong password policies, enforcing complexity and regular changes. Keep your WAP firmware updated to patch known vulnerabilities; manufacturers frequently release updates addressing newly discovered security flaws. This seemingly simple practice can significantly improve your security posture, preventing exploitation of known weaknesses.

Regularly back up your WAP configuration. This safeguard allows you to quickly restore your system to a secure state in case of compromise or accidental misconfiguration. It ensures business continuity and minimizes downtime in the event of a security incident, acting as an insurance policy against unexpected disruptions. Moreover, consider implementing a robust change management process for any configuration changes on your WAPs. This controlled approach helps prevent accidental misconfigurations that could introduce new security vulnerabilities.

Furthermore, segment your network using VLANs to isolate sensitive devices and data from less critical ones. This limits the impact of a breach, preventing attackers from accessing your entire network if they compromise a single WAP. Employ strong encryption protocols, such as WPA3, to protect wireless communications. These protocols use robust algorithms to secure data transmitted between devices and the WAP, rendering interception much more difficult.

Finally, educate your users about basic security best practices. This includes encouraging them to use strong passwords, avoiding public Wi-Fi networks when possible, and being wary of phishing attempts. Remember, even the most secure WAP is vulnerable if your users fall prey to social engineering tactics. Regular training keeps your users informed and vigilant, providing a crucial layer of defense against sophisticated attacks, significantly reducing human error related breaches. A combination of technical measures and user awareness forms a comprehensive and highly effective approach to security. Therefore, a comprehensive security strategy requires a multi-faceted approach, encompassing both technological defenses and user education. The regular execution of these security practices will ensure a more robust and resilient wireless network.

6.1 Vulnerability Assessments

Regular vulnerability assessments are critical for maintaining the security of your wireless access points (WAPs). They provide a snapshot of your current security posture, revealing weaknesses before they can be exploited by malicious actors. Think of it as a thorough health check for your network's most exposed point. A proactive approach is far more cost-effective than reacting to a breach.

Effective vulnerability assessments begin with identifying all your WAPs. This might seem simple, but many organizations overlook forgotten or poorly documented access points, creating significant security gaps. Once identified, you need to categorize them based on their criticality—a WAP in a high-security area like the server room demands far more attention than one in a less sensitive location. This prioritization streamlines your assessment process and focuses your efforts on the most vulnerable assets.

Next, utilize automated vulnerability scanners. These tools can quickly scan your WAPs for known vulnerabilities, providing a comprehensive report detailing identified weaknesses. Remember to choose a reputable scanner that is regularly updated with the latest vulnerability databases; outdated tools offer a limited view and may miss critical threats. These automated scans are just the starting point, they should be supplemented with more rigorous manual checks, where needed.

Manual assessments are crucial for discovering vulnerabilities that automated scans often miss. This could involve analyzing network traffic patterns for suspicious activity, checking for weak passwords or default configurations, and examining the WAP's firmware for known exploits. Don't forget to consider physical security, too. An unsecured WAP is far easier to compromise than one protected by physical barriers and access control.

Following the assessment phase, meticulously document all findings. Create a detailed report outlining each identified vulnerability, its severity, and the potential impact of exploitation. This report will be your roadmap for remediation. Prioritize vulnerabilities based on their severity and likelihood of exploitation. Addressing high-severity, high-likelihood vulnerabilities should be your immediate focus.

Once you've identified and prioritized your vulnerabilities, develop a comprehensive remediation plan. This isn't just about patching software; it involves configuring firewalls, implementing access controls, strengthening passwords, and educating users about security best practices. Remember that effective security is a continuous process; it requires consistent monitoring and updates to remain effective in the face of evolving threats.

Don't underestimate the value of regularly scheduled vulnerability assessments. Treat them as a vital part of your overall security strategy. The frequency will depend on your network's complexity and risk tolerance, but quarterly or even monthly scans are recommended for high-security environments. Regular assessments ensure that you remain ahead of potential threats and minimize your exposure.

Finally, consider involving a third-party security consultant for periodic audits. An independent assessment can provide an objective perspective, uncovering weaknesses that your internal team might miss. A fresh set of eyes can often identify vulnerabilities and inefficiencies that have become ingrained within an organization over time. While it adds cost, the potential benefits of avoiding a costly breach far outweighs the expense. They can bring expertise in cutting-edge attack vectors and security protocols that keeps you up-to-date with the latest threats. The expertise in specialized tools and techniques enhances the effectiveness and comprehensiveness of the assessment, leading to more robust security measures. Remember this is not an optional expenditure for high-risk environments, but an essential investment for business continuity and data protection. A professional assessment also offers the advantage of a documented, verifiable assessment to demonstrate compliance with industry regulations and best practices.

Investing the time and resources in regular vulnerability assessments is not just about meeting compliance requirements; it's about safeguarding your business from significant financial and reputational damage. By proactively identifying and addressing weaknesses, you build a robust defense against the ever-evolving landscape of cyber threats. The peace of mind that comes with knowing your WAPs are secure is invaluable. Remember, a comprehensive approach, combining automated scans, manual checks, detailed documentation, and a well-defined remediation plan, is the key to maintaining strong WAP security. This consistent vigilance is the cornerstone of a robust security strategy, ensuring the confidentiality, integrity, and availability of your sensitive data. Don't wait for a breach to occur; be proactive and protect your network. Treat vulnerability assessments as an ongoing investment in the safety and security of your valuable data and your organizational reputation.

6.2 Penetration Testing

Penetration testing isn't just about breaking things; it's about understanding vulnerabilities before attackers do. It's a proactive, ethical hacking approach that simulates real-world attacks to identify weaknesses in your WAP security infrastructure. Think of it as a comprehensive security checkup, revealing hidden flaws that could leave your network exposed. This methodical approach allows for preemptive patching and strengthening of defenses, minimizing the impact of future breaches. The goal isn't just finding problems, but also providing solutions.

Before you launch into a penetration test, meticulous planning is crucial. Clearly define the scope of your test—which WAPs, what types of attacks to simulate, and what level of access the testers will have. This prevents unintended damage or disruptions to your network. Detailed documentation of the testing methodology, including tools and techniques employed, safeguards against misunderstandings. A well-defined scope also ensures efficient allocation of resources and time, focusing efforts on the most critical aspects of the WAP infrastructure. Remember, a well-planned test is the foundation for effective results.

Choosing the right tools is vital for a successful penetration test. Network scanners identify active devices and open ports, providing insights into potential entry points. Packet sniffers capture network traffic, allowing analysis of communication patterns and potential vulnerabilities. Wireless intrusion detection systems can uncover unauthorized access attempts. Ethical hacking frameworks like Metasploit provide a suite of tools for simulating various attacks, offering a range of approaches to assess the robustness of your WAP security. However, remember that selecting the right tools is only part of the equation; skillful use and careful interpretation of results are paramount.

The testing process itself usually starts with reconnaissance, essentially gathering information about the target WAP. This involves passively monitoring network traffic, identifying the WAP's make and model, and searching for publicly available information about the network. Next comes the vulnerability scan; automated tools are usually employed here to identify known weaknesses in the WAP's firmware, configuration settings, or even connected devices. Following that, an active penetration test involves attempting to exploit the discovered vulnerabilities, testing the resilience of the security measures. This might include attempts to crack passwords, inject malicious code, or gain unauthorized access to the network. Finally, a thorough report documents the findings, detailing the discovered vulnerabilities, their severity, and recommended remediation steps. This detailed report is pivotal for guiding the subsequent strengthening of your WAP security posture.

Following the penetration test, the remediation phase begins, and this requires both technical expertise and careful project management. Each discovered vulnerability needs a corresponding solution. This could involve patching software, reconfiguring network settings, implementing stronger authentication mechanisms, or even replacing outdated equipment. Prioritize critical vulnerabilities first, focusing on those that pose the greatest threat to your network. The implementation of these fixes must be meticulously documented to provide a comprehensive record of the improvements made. A post-remediation test validates the effectiveness of the changes implemented; this is an essential step to ensure the vulnerabilities have been successfully addressed. This iterative process of testing, remediation, and retesting ensures a continuous improvement of the overall security posture.

Effective penetration testing requires a blend of technical skills and a deep understanding of network security principles. Testers need expertise in network protocols, encryption techniques, and common attack methods. However, soft skills are equally crucial; communication, teamwork, and the ability to clearly articulate findings are essential for successful collaboration with other stakeholders. This holistic approach integrates technical expertise with effective communication, leading to better implementation of recommendations and improving overall security measures. Ultimately, the success of penetration testing lies not only in identifying vulnerabilities but also in facilitating their effective remediation.

Remember, the goal is not to merely find flaws, but to enhance the security posture of your entire WAP infrastructure. Regular penetration testing, combined with ongoing security monitoring and proactive updates, forms a robust defense against evolving threats. By continually evaluating and refining your security protocols, you minimize your risk and protect your network's integrity and the confidentiality of your data. Consider engaging professional penetration testers who bring specialized skills and experience to ensure a comprehensive and effective evaluation. Their expertise enhances the security of your network, ensuring a robust defense against cyber threats. Proactive security measures, like penetration testing, significantly reduce the potential damage from successful attacks. It is a cost-effective investment that safeguards your valuable data and network infrastructure.

7 Future of WAP Security

The integration of artificial intelligence and machine learning will revolutionize WAP security. AI algorithms can analyze massive datasets of network traffic, identifying anomalous patterns indicative of malicious activity far more efficiently than traditional signature-based systems. This proactive approach enables real-time threat detection and response, significantly reducing the window of vulnerability. Machine learning models can adapt and learn from new attack patterns, providing ongoing protection against evolving threats. This dynamic capability is crucial in today's landscape where attackers constantly refine their tactics. The implementation of AI-driven security solutions will necessitate robust data processing infrastructure and skilled personnel capable of interpreting AI-generated insights. Furthermore, careful consideration must be given to the potential for AI bias and the need for transparency in AI-driven decision-making. The ethical implications of utilizing AI in security are paramount and warrant ongoing discussion and refinement of best practices. Ultimately, the integration of AI promises a significant leap forward in the effectiveness of WAP security, but responsible implementation is critical to its success.

Next-generation encryption techniques will play a pivotal role in the future of WAP security. Post-quantum cryptography, designed to withstand attacks from quantum computers, is of critical importance. The development and implementation of these algorithms, such as lattice-based cryptography and code-based cryptography, are crucial to securing long-term confidentiality. While classical encryption algorithms like AES remain effective, their eventual vulnerability to quantum computing necessitates a proactive transition to quantum-resistant alternatives. This transition requires careful planning and coordination across the industry, ensuring compatibility and minimizing disruption to existing infrastructure. The standardization and adoption of post-quantum cryptography is a complex undertaking, requiring collaboration between researchers, developers, and standardization bodies. The effective integration of these next-generation encryption methods will represent a critical step in assuring the long-term security of WAP networks.

Securely integrating the Internet of Things (IoT) devices into WAP networks poses significant challenges. The sheer volume and diversity of IoT devices, often with limited processing power and security features, create a large attack surface. Robust authentication and authorization mechanisms are paramount, as are secure communication protocols tailored for resource-constrained IoT devices. Implementing a secure device management system, allowing for remote updates and configuration changes, is crucial for mitigating vulnerabilities and ensuring ongoing security. The development of lightweight cryptography algorithms, specifically designed for IoT devices, will be essential to maintaining performance while securing communications. Furthermore, the use of edge computing to process sensitive data locally reduces the reliance on bandwidth and lowers the risk of data breaches during transmission. The successful integration of IoT security within WAP architecture demands a holistic approach, encompassing device security, network security, and data management strategies.

Zero Trust architectures fundamentally shift the paradigm of network security. Instead of assuming trust within the network perimeter, Zero Trust models verify every user and device attempting access to resources, regardless of location. This micro-segmentation approach limits the impact of potential breaches, isolating compromised components and preventing lateral movement within the network. Continuous authentication and authorization, leveraging multi-factor authentication and identity management systems, are crucial components of a Zero Trust approach. This continuous verification ensures that only authorized users and devices can access sensitive data and functionalities. The implementation of a robust Zero Trust architecture demands careful planning and integration with existing network infrastructure. It necessitates a shift in security mindset and the adoption of new technologies, including access control policies and continuous monitoring tools, to effectively manage and protect the dynamic nature of modern WAP networks. Successful adoption relies on comprehensive security training for both users and administrators.

The convergence of these advancements will shape the future of WAP security. The interplay between AI-driven threat detection, next-generation encryption, secure IoT integration, and Zero Trust architectures will create a more robust and resilient wireless network environment. However, the successful adoption of these technologies demands careful planning, proactive investment, and ongoing adaptation to evolving threats. A holistic approach, integrating security into every aspect of WAP network design and management, is crucial for ensuring a secure and reliable wireless infrastructure in the future. Continuous monitoring, regular security assessments, and prompt updates will be essential to effectively maintain the security posture of WAP networks in this ever-changing landscape. The proactive adoption of these advancements is not just about technological upgrades; it is about fostering a security-conscious culture, proactively addressing emerging challenges, and building a more secure digital future.

7.1 AI & Machine Learning

AI and machine learning are rapidly transforming cybersecurity, offering promising solutions for enhancing WAP security. Their ability to analyze vast amounts of network data in real-time allows for the detection of subtle anomalies often missed by traditional methods. This proactive approach enables faster response times and more effective mitigation strategies. For instance, machine learning algorithms can identify unusual traffic patterns indicative of an ongoing attack, even before a significant breach occurs. This surpasses traditional signature-based detection systems which rely on predefined attack signatures and are often unable to adapt quickly enough to novel threats. We can expect improved accuracy and speed in identifying threats.

The implementation of AI and machine learning in WAP security isn't merely theoretical; it's already yielding tangible benefits. Many vendors are incorporating these technologies into their network security products, offering features like automated threat detection, predictive analysis, and adaptive security measures. These products learn and adapt as they process data, improving their efficacy over time. This adaptive capability is crucial in a constantly evolving threat landscape where attackers consistently develop new tactics and techniques. Imagine an AI-powered system that not only detects an attack but also automatically adjusts the firewall rules to block future attempts from the same source, adapting its response based on the learning process, preventing repeat incidents and reducing response time, preventing a potentially large-scale data breach.

However, integrating AI and machine learning into WAP security presents certain challenges. The sheer volume of data generated by a typical network can be overwhelming, requiring significant computational resources for effective analysis. Data quality also poses a problem; inaccurate or incomplete data can lead to false positives or false negatives, resulting in unnecessary alerts or missed attacks respectively. Furthermore, the algorithms themselves require meticulous design, testing, and ongoing refinement to ensure accuracy and prevent adversarial attacks. Adversarial examples, specifically crafted data designed to deceive AI systems, can bypass even the most sophisticated algorithms, creating a weakness in the overall security system. Thus, rigorous testing and validation are crucial to ensure the reliability and trustworthiness of these systems.

Effective implementation requires a multifaceted approach. Begin by focusing on data quality. Ensure your network is instrumented to collect comprehensive and accurate data. This means implementing robust logging mechanisms and properly configuring your security devices. Second, select AI and machine learning tools carefully, considering factors such as scalability, accuracy, and the ability to integrate with existing security infrastructure. Choosing the right tools is just as critical as the implementation itself. Choosing vendor products may mean incorporating legacy systems which may be less efficient or not adaptable. Third, prioritize ongoing monitoring and evaluation. Regularly assess the performance of your AI-powered security systems, adjusting parameters and retraining models as needed. AI and ML models are not static, and require constant monitoring and refinement to remain effective in tackling evolving threats. This ensures that the systems remain effective and responsive to the changing threat landscape, maintaining their efficacy over time.

Finally, consider the ethical implications. AI-powered security systems have the potential to be biased, leading to unfair or discriminatory outcomes. It's crucial to implement safeguards to mitigate these risks, focusing on fairness and transparency. These systems should be carefully reviewed for bias in their algorithms and their data inputs to ensure that they do not discriminate against certain users or groups. This ensures fair and equitable implementation of the AI and ML model, preventing unjust outcomes. By proactively addressing these challenges, organizations can harness the full potential of AI and machine learning to significantly enhance WAP security. The integration of these technologies is not merely an add-on but a significant shift in the approach to network protection, offering a new layer of security. Ultimately, a robust and well-integrated AI-powered security system can provide a significantly enhanced level of protection for your wireless network infrastructure, mitigating risks and safeguarding valuable data.

7.2 Next-Gen Encryption

Post-quantum cryptography is emerging as a crucial area within next-generation encryption. The threat of quantum computers breaking current encryption standards is a very real and imminent one. Organizations must begin planning for a transition now, instead of waiting for a crisis to force their hand. Failing to do so risks significant data breaches and financial losses once quantum computers achieve sufficient processing power. This isn't about hypothetical future threats; this is about proactive risk mitigation.

Lattice-based cryptography offers strong security against both classical and quantum computers. It's a relatively new field, but significant advancements have already been made. Algorithms like Kyber, Crystals-Dilithium, and Falcon have been standardized by NIST and are showing great promise for practical applications. Unlike older encryption methods, these algorithms are designed specifically to withstand the computational power of a quantum computer. This robust design is absolutely critical for the future of data security.

Implementing lattice-based cryptography isn't a simple switch. It demands a fundamental shift in infrastructure. Existing systems often rely on RSA and ECC, making direct replacement difficult. Therefore, a phased approach is strongly recommended, beginning with a comprehensive assessment of existing security infrastructure and identifying critical systems requiring immediate upgrades. This is not just a technology update, it's a strategic restructuring of security protocols.

Furthermore, the transition needs a strong focus on interoperability. Adopting new encryption algorithms requires careful consideration of how they will integrate with existing hardware and software. Open standards are crucial to ensure compatibility and avoid vendor lock-in, allowing for easier migration and preventing fragmentation in the ecosystem. Interoperability guarantees seamless deployment and avoids costly compatibility issues in the long run. It's a key factor in smooth implementation.

Beyond specific algorithms, the move to next-generation encryption necessitates improved key management practices. Keys are the heart of any encryption system, and securing them is paramount. Robust key generation, storage, and rotation procedures are essential. Implementing Hardware Security Modules (HSMs) is highly advised, especially for storing private keys. HSMs provide a physically secure environment to protect cryptographic keys from theft or unauthorized access. Think of it as the ultimate vault for your digital assets.

Additionally, the rise of the Internet of Things (IoT) presents unique challenges. Many IoT devices have limited processing power and memory, making implementation of complex post-quantum algorithms difficult. Lightweight cryptography is thus becoming vital. This requires research and development of encryption schemes specifically optimized for resource-constrained environments. The solutions must be efficient and secure; otherwise, the increased security is void.

Education and training are also essential components of the transition to next-generation encryption. Security teams need to understand the implications of quantum computing and the capabilities of post-quantum algorithms. This involves familiarization with the mathematical principles behind the new algorithms, as well as hands-on experience with implementing and managing these systems. Effective training will ensure seamless operation and problem resolution. It is a proactive, not reactive strategy.

Finally, staying informed about the ever-evolving landscape of post-quantum cryptography is crucial. New algorithms and research are constantly emerging, and organizations must remain adaptable. Following the work of leading cryptographic researchers and industry organizations provides insight into the future of encryption. Active monitoring of standards and best practices is vital for maintaining a strong security posture in a constantly evolving digital world. This involves regular updates and consistent vigilance.

7.3 IoT Security Integration

The explosive growth of the Internet of Things (IoT) presents both incredible opportunities and significant security challenges for wireless access points (WAPs). Integrating IoT devices securely into existing WAP infrastructures requires a multi-faceted approach, demanding careful consideration of device authentication, data encryption, and network segmentation. Failure to address these aspects leaves networks vulnerable to a wide array of attacks, from simple data theft to sophisticated, coordinated intrusions. This necessitates a proactive, layered security strategy.

Effective IoT device authentication is paramount. Unlike traditional devices, IoT devices often have limited processing power and memory, making standard authentication methods impractical. Lightweight protocols, such as MQTT with appropriate TLS encryption, are better suited for these devices, minimizing the computational overhead while maintaining a reasonable security posture. However, relying solely on device authentication is insufficient.

Securing the communication channels between IoT devices and the WAP is crucial. Data-in-transit needs robust encryption, ideally using AES-256 with perfect forward secrecy (PFS). PFS ensures that even if a cryptographic key is compromised, past communication sessions remain confidential. The use of Transport Layer Security (TLS) or Datagram Transport Layer Security (DTLS) is essential, depending on the communication protocol employed. Regular updates of cryptographic libraries and protocols are necessary to mitigate vulnerabilities discovered in existing implementations. Regular audits are essential to ensure consistent application of these security measures. Network segmentation plays a pivotal role in mitigating the impact of a security breach. IoT devices should be isolated from the rest of the network through virtual LANs (VLANs) or other segmentation techniques. This limits the potential damage if an IoT device is compromised; attackers are prevented from easily accessing sensitive information within the core network. Firewalls, properly configured to permit only necessary communication between IoT devices and other network segments, are another essential layer of defense. The complexity of managing these firewalls warrants skilled personnel and robust management protocols.

Effective monitoring and incident response are crucial elements of any comprehensive security strategy for IoT-integrated WAPs. Security Information and Event Management (SIEM) systems provide a centralized platform to collect and analyze security logs from various sources, including IoT devices, the WAP, and other network components. This allows for timely detection of anomalies and potential security incidents, facilitating swift and effective response. The SIEM system's capabilities should be tailored to the specific types of threats and vulnerabilities present in the IoT network. Furthermore, comprehensive logging across all connected devices and network infrastructure is essential for thorough incident analysis and future prevention.

Developing a robust security strategy for IoT-integrated WAPs requires a holistic approach. It goes beyond simply deploying security technologies; it necessitates a change in mindset, shifting from reactive to proactive security measures. This includes developing comprehensive security policies, conducting regular security audits and penetration testing to identify vulnerabilities, and establishing strong incident response plans. Only through careful planning, robust implementation and continuous monitoring can organizations adequately address the growing security challenges presented by the integration of IoT devices into their wireless networks. Regular staff training is vital to ensure consistent application of security practices and the prompt detection of anomalies.

The ongoing evolution of IoT technology necessitates a flexible and adaptable security strategy. New threats constantly emerge, demanding continuous adaptation of security protocols, technologies, and practices. Staying abreast of the latest security vulnerabilities and best practices through continuous professional development is crucial for network administrators. Understanding these complexities and implementing appropriate security measures requires a deep understanding of both IoT technologies and network security principles. Ignoring these precautions leaves the network vulnerable to potential data breaches, service disruption, and reputational damage. Only through diligent application of best practices can organizations derive the full benefits of IoT technology while minimizing the risks. Effective security is not an end point, but a continuous process of improvement and adaptation.

7.4 Quantum-Resistant Cryptography

Quantum computing poses a significant threat to current encryption standards. Algorithms that are computationally infeasible to crack with today's classical computers become vulnerable to the immense processing power of quantum computers. This necessitates a proactive shift towards quantum-resistant cryptography, safeguarding our networks from future attacks. This transition requires careful planning and a phased approach, not a sudden switch.

The core challenge lies in developing algorithms that remain secure even against quantum attacks. While current encryption methods like AES rely on the difficulty of factoring large numbers or solving discrete logarithm problems—tasks easily solved by quantum computers—post-quantum cryptography explores alternative mathematical problems believed to be intractable for both classical and quantum computers. This isn't a simple swap; it demands a deep understanding of the underlying mathematics and the practical implications of implementation.

Lattice-based cryptography, for instance, offers a promising avenue. It relies on the hardness of finding short vectors in high-dimensional lattices. This problem remains computationally challenging even for quantum computers, offering robust security. Code-based cryptography, another contender, leverages the difficulty of decoding random linear codes. These systems, while potentially complex to implement, provide another strong layer of defense against quantum adversaries. Understanding the strengths and weaknesses of each is critical for effective deployment.

Implementing quantum-resistant cryptography isn't merely about swapping algorithms; it demands a comprehensive review of existing infrastructure. Network devices, software, and protocols all need to be assessed for compatibility and updated accordingly. This isn't just a technology upgrade; it's a strategic overhaul of security architecture. A phased migration strategy, starting with high-value assets and gradually expanding to encompass the entire network, minimizes disruption while ensuring ongoing protection.

The migration process will undoubtedly present challenges. The performance overhead of some post-quantum algorithms might be substantial compared to their classical counterparts. Careful consideration is necessary to avoid unacceptable latency or throughput issues. Furthermore, the ongoing research into quantum-resistant cryptography means that the landscape is still evolving. Staying informed about the latest advancements and best practices is essential for maintaining a strong security posture.

Interoperability is another critical aspect. Different quantum-resistant algorithms might not be compatible with each other. Therefore, standardization efforts play a crucial role in ensuring seamless integration across diverse systems and networks. The selection of the most appropriate algorithm will depend on the specific needs of the network, factoring in factors like performance, security requirements, and compatibility with existing infrastructure. This careful selection process is crucial for a smooth and effective migration.

The development of robust key management practices for post-quantum cryptography is equally paramount. Existing key management systems might need upgrades to accommodate the new algorithms and ensure secure key generation, storage, and distribution. This requires a significant investment in infrastructure and expertise, emphasizing the need for long-term planning and resource allocation. Failing to address this aspect could jeopardize the overall security offered by quantum-resistant cryptography.

Finally, consider the human factor. The successful transition to post-quantum cryptography hinges not just on technology, but also on expertise and training. Network administrators and security personnel need to understand the principles of post-quantum cryptography, its implications, and best practices for implementation and management. Investing in comprehensive training programs and awareness initiatives is as crucial as the technology itself, ensuring a secure and efficient transition. A well-trained team can effectively implement and manage the new systems, guaranteeing the continued protection of the network. This investment in human capital should be regarded as integral to the overall success of this critical change.

7.5 Zero Trust Architectures

Zero Trust, at its core, shifts the paradigm from implicit trust to explicit verification. Instead of assuming that anything inside your network is inherently safe, a Zero Trust architecture treats every access request—regardless of origin—as potentially hostile. This fundamental shift requires a complete reassessment of your security posture, demanding granular control over access and rigorous authentication at every point. It's not just a technology upgrade; it's a fundamental change in philosophy.

Implementing Zero Trust isn't a single, easily applied solution. It's an iterative process involving multiple layers. Imagine building a castle, not with a single, easily breached wall, but with multiple concentric rings of defenses. Each layer adds another hurdle for an attacker to overcome, dramatically increasing the difficulty and cost of a successful intrusion. This multi-layered approach is crucial for a truly effective Zero Trust implementation.

For a beginner, the most accessible starting point lies in strong authentication. This includes moving beyond simple passwords to multi-factor authentication (MFA), ideally using a combination of something you know (password), something you have (phone or security token), and something you are (biometrics). Implementing MFA across all access points significantly raises the bar for attackers. It's a relatively straightforward step with a significant impact on security. Failure to enforce MFA leaves your network vulnerable, no matter how sophisticated the rest of your security measures.

Next, consider micro-segmentation. Instead of treating your entire network as a single entity, break it down into smaller, isolated segments. This limits the damage of a successful breach; if an attacker gains access to one segment, they won't automatically have access to your entire infrastructure. Think of it as containing a fire within a specific room, preventing it from spreading throughout the building. Effective micro-segmentation requires careful planning and implementation, potentially involving VLANs and carefully configured firewalls.

Data security is paramount within a Zero Trust framework. Every piece of data should be classified and protected according to its sensitivity. This may involve encryption both in transit and at rest, using robust encryption algorithms like AES-256. Regularly review and update your data access policies; ensure that only authorized users have the necessary access permissions, based on the principle of least privilege. This involves continuous monitoring and adjustment based on the ever-changing needs of your organization.

Beyond technical measures, a robust Zero Trust architecture needs a strong human element. Comprehensive security awareness training for all users is critical. Employees need to understand the importance of strong passwords, the dangers of phishing attacks, and the significance of reporting suspicious activity. Regular security audits and penetration testing help identify and remediate vulnerabilities before attackers can exploit them. Remember, your security is only as strong as its weakest link, and often that link is human error. This continual education and awareness are non-negotiable.

Finally, embracing a Zero Trust model often necessitates adopting a more proactive security strategy. This involves deploying Intrusion Detection and Prevention Systems (IDS/IPS) to monitor network traffic for suspicious activity and react to potential threats in real-time. It also requires robust Security Information and Event Management (SIEM) capabilities to gather, analyze, and correlate security logs from various sources, allowing you to quickly detect and respond to incidents. Regularly updating your security tools and infrastructure is as crucial as the initial implementation. The cyber threat landscape is ever-evolving, demanding constant adaptation and vigilance.

Implementing a Zero Trust architecture is a journey, not a destination. It demands continuous monitoring, adaptation, and a commitment to improving your security posture over time. It's a complex undertaking that requires careful planning, consistent implementation, and continuous improvement. Begin with strong authentication, proceed to micro-segmentation, prioritize data security, invest in user training, and maintain constant vigilance. This is how to build a secure and resilient network in today's ever-evolving threat landscape. Remember, a well-implemented Zero Trust framework isn't just about technology; it's about a fundamental shift in how you approach network security.

www.ingramcontent.com/pod-product-compliance
Lightning Source LLC
LaVergne TN
LVHW022021200926
848481LV00007B/2466
9798348555238